Monitoring of Neurological Parameters in Critical Care Settings

Editor /Author

Salman Assad, M.D.

Shifa Tameer-e-Millat University, Islamabad, Pakistan

About this Book

This book of provides a complete and comprehensive overview of modern management of neurologic monitors in ICUs and CICUs in all its complexity, from basic science to gold-standard care.

- Beautifully produced in full color throughout, and with high-quality illustrations, it successfully:
- Provides a solid overview of what clinicians/surgeons/anesthesiologists do, and with topics presented in an order that a clinician will encounter them.
- Presents a holistic look at immunologic responses, foregrounding the interrelationships between team members.

Table of Contents

Neuromonitoring in Cardiac Intensive Care Unit (CICU)

Salman Assad, MD

Department of Neurology and Neurosurgery

Shifa Tameer-e-Millat University, Islamabad, Pakistan

Introduction

Critically ill patients in cardiac intensive care unit (CICU) often have possibility of neurological dysfunction either as anend result of primary or secondary neurological insults. In recent years, various neurologic monitoring techniques have evolved and now play a significant role in management of patients with brain injury. Combination of monitoring techniques is required due to presence of different variables in each patient.

1. Neurological Assessment

Following are the components for neurological assessment;

1. Interview
2. Level of Consciousness
3. Pupillary Assessment
4. Cranial Nerve Testing
5. Vital signs
6. Motor Function
7. Sensory Function
8. Tone
9. Cerebral Function

2. Intracranial pressure (ICP) monitoring

The fundamental principle of intracranial pressure (ICP) is based on the *Monro-Kellie doctrine* that here is a fixed volume within the enclosed skull that determines the pressure. George Burrows suggested in 1846 the idea of a reciprocal relationship between the volumes of CSF and blood. Harvey Cushing in 1926 proposed that within undamaged skull, the volume of the brain, blood, and CSF is constant. An increase in one component will cause a decrease in one or both of the other components.

Indications for ICP Monitoring

Cushing's triad (hypertension, bradycardia and apnea) is an indicator of intracranial hypertension. An approach to deciding which disorders should be evaluated with intracranial pressure monitors is based on the conditions which typically result in ICP elevations as shown in table below.

Table: Conditions where ICP-monitoring is used

- Traumatic head injury
- Intracerebral hemorrhage
- Subarachnoid hemorrhage
- Hydrocephalus
- Malignant infarction
- Cerebral edema
- CNS infections
- Hepatic encephalopathy

The duration of monitoring is until ICP has been normal for 24 to 48 hours without ICP therapy.

Table: Indications for ICP monitoring

ICP Monitoring
1. GCS Score: 3–8 (after cardiopulmonary resuscitation)

2. Abnormal Admission Head CT Scan
  a. Hematoma
  b. Contusion
  c. Edema
  d. Herniation
  e. Compressed basal cisterns

3. Normal Admission Head CT Scan PLUS 2 or more of the following
  a. Age > 40 years
  b. Motor posturing (decerebrate or decorticate)
  c. Systolic blood pressure < 90 mm Hg

Normal range of ICP

Variations of normal ICP are based on age and body posture. Normal ICP is generally in between 5–15 mmHg in healthy supine adults, 3–7 mmHg in children and 1.5–6 mmHg in infants. Raised ICP can lead to brain herniation.

Monitoring Devices for ICP

Most common and well-known methods assess whether noninvasive techniques (transcranial Doppler, optic nerve sheath diameter,CT scan/MRI , fundoscopy and tympanic membrane displacement) can be used as reliable alternatives to the invasive techniques (ventriculostomy and micro transducers). Ventriculostomy with external ventricular drainage (EVD) is considered the gold standard in terms of accurate measurement of pressure. Lesser complications of hemorrhage and infections are encountered in non-invasive techniques as compared to invasive techniques.

Table: Invasive and Non-invasive Methods of ICP Measurement

ICP Measurement

Non-Invasive Methods

- Transcranial Doppler (TCD) Ultrasonography
- Optic Nerve Sheath Diameter
- Fundoscopy
- Typmanic Membrane Displacement (TMD)

Invasive Methods

- External Ventricular Drainage (EVD)
- Micro-transducer ICP Monitoring Device

4. Electroencephalography (EEG) monitoring in ICU

Indications

Few important indications for EEG monitoring in ICU;

- Seizures

- Status Epilepticus (SE) and Non-convulsive status epilepticus (NCSE)

- Metabolic Enccepahlopathy

- Hypoxic Ischemic Encephalopathy

- Early detection of ischemia in patients with subarachnoid hemorrhage at higher risk of vasospasm

Status Epilepticus (SE)

Status epilepticus (SE) is defined as 5 min or more of (a) continuous clinical and/or electrographic or (b) recurrent seizure activity without recovery (returning to baseline) between seizures.

Nonconvulsive SE (NCSE)

Nonconvulsive SE (NCSE) is most commonly observed in critically ill patients in intensive care setting with severely impaired/altered mental status, with or without subtle motor movements (e.g., rhythmic muscle twitches or tonic eye deviation that often occurs during acute brain injury). NCSE in the ICU frequently follows uncontrolled or partially treated generalized convulsive SE (GCSE).

Continuous EEG (cEEG) monitoring (over 24 h) is needed to diagnose NCSE and to manage refractory SE. cEEG monitoring should be initiated within 1 h of SE onset if ongoing seizures are suspected in all patients. cEEG monitoring should be done for at least 48 h following acute brain insult in comatose patients to evaluate non-convulsive seizures and 24 h after the cessation of electrographic seizures or during antiepileptic drug (AED) weaning trials.

5. Trans-cranial Doppler (TCD) Role in Neuro-monitoring

Trans-cranial Doppler (TCD)ultrasonography is an inexpensive, non-invasive, portable and safe technique that uses a pulsed Doppler transducer for assessment of cerebral blood flow velocity (CBFV) in ICU setting.

Technique:

TCD is performed at the bedside. The first step is to localize a cranial "window" where the ultrasound beam can penetrate without being excessively dampened. The three main windows for accessing the intracranial arteries are listed below.

Transtemporal window—found between the angle of the eye and the pinna above the zygomatic ridge and is the major route for insonating the anterior, middle and posterior cerebral arteries.

Transorbital window—through the eye for insonation of the ophthalmic artery and the siphon of the internal carotid artery.

Transforaminal window—through the foramen magnum insonated from the top of the neck below the occiput for the basilar artery and the intracranial segments of the vertebral arteries.

Table: Uses of transcranial Doppler (TCD) ultrasound

- Detection of site/degree of stenosis/occlusion of cerebral vasculature
- Assessment of recanalisation following occlusion (with/without thrombolytic treatment)
- Assessment of collateral flow in intracranial vasculature in cases of critical carotid artery stenosis (extracranial)
- Detection of microemboli: stratification of risk of recurrence of stroke/TIA
- Detection and quantification of right to left shunts
- Detection of degree of vasospasm following subarachnoid haemorrhage
- Complementary to duplex carotid scan in diagnosis of subclavian steal syndrome
- Intra-operative monitoring of carotid endarterectomy

6. Somatosensory Evoked Potentials (SSEPs)

The somatosensory evoked potentials (SSEPs) monitoring is the electrophysiological response of the nervous system to sensory stimulation. The peripheral mixed nerves are stimulated electrically, and the response is measured along the sensory pathway. SSEP reflects the functional integrity of the somatosensory pathways.

Indications for SSEPs

Following are the indications of SSEPs;

Surgeries of the spine

- Correction of scoliosis with instrumentation
- Spinal cord decompression and stabilization after acute spinal cord injury
- Spinal fusion
- Release of tethered cord
- Resection of spinal cord tumor/cyst/vascular lesion
- Correction of cervical spondylosis

Surgeries of the brain

- Localization of the sensorimotor cortex
- Clipping of intracranial aneurysms
- Resection of intracranial vascular lesions involving the sensory cortex and arteriovenous malformation
- Resection of thalamic tumor
- Brainstem surgeries

Vascular surgery

- Carotid endarterectomy (CEA)

- Abdominal and thoracic aortic aneurysm repair

- Repair of coarctation of the aorta

Intensive Care Unit

- Prognostication in hypoxic-ischemic encephalopathy/traumatic brain injury (TBI)/post-cardiac arrest

- Acute neurological deteriorations

Others

- Brachial plexus exploration after injury

- Positioning during surgeries of the skull base

- Chronic pain

- Herpetic neuralgia

Guide to prognosis

Clinical assessment of the comatose ICU patient is limited to examination of the brainstem reflexes and motor responses. SSEPs can be used to enhance prognostic predictions in post-traumatic and anoxic–ischemic coma. They are less susceptible to the effects of metabolic changes and sedating agents than clinical signs, such as motor responses and to a lesser extent the pupillary light reactions. Meta-analyses of the bilateral absence of cortical N20 responses, recorded after 72 hours, can predict death or the persistent vegetative state (that is, non-awakening) with a specificity of > 99% in anoxic–ischemic and around 95% in traumatic coma.

Prognostication after Cardiac Arrest

i. Role of EEG in prognostication after Cardiac Arrest

EEG is an essential part of prognostication after cardiac arrest but precise role of EEG in predicting and improving outcomes in survivors of cardiac arrest (CA) is still under discussion. In a recent prospective study Rossetti et al., the best estimation of poor long-term neurologic outcome in survivors of cardiac arrest (CA) treated with therapeutic hypothermia (TH) was accomplished with the presence of any two of the following four independent outcome predictors:

- Nonreactive EEG background,
- Incomplete recovery of brainstem reflexes,
- Bilaterally absent Somato-sensory Evoked Potentials (SSEP)s
- Myoclonus

ii. Role of SSEPs in prognostication after Cardiac Arrest

In patients with cardiac arrest, the median nerve is most commonly stimulated bilaterally at the wrist. Electrodes are then placed at the elbow, Erb's point, the cervical medulla (peripheral) and on the parietal and frontal cortex (cortical). If peripheral responses are not present, this may be due to peripheral nerve damage. For prognosis of a poor outcome after CA, only the short cortical latencies (N20, expected to appear 20 milliseconds after median nerve stimulation) are used. In order to have absent SSEPs, predictive of a poor outcome, cortical responses have to be absent bilaterally in a technically well-performed test. *In patients who remain comatose after CA, SSEPs have been shown to reliably predict poor outcome.*

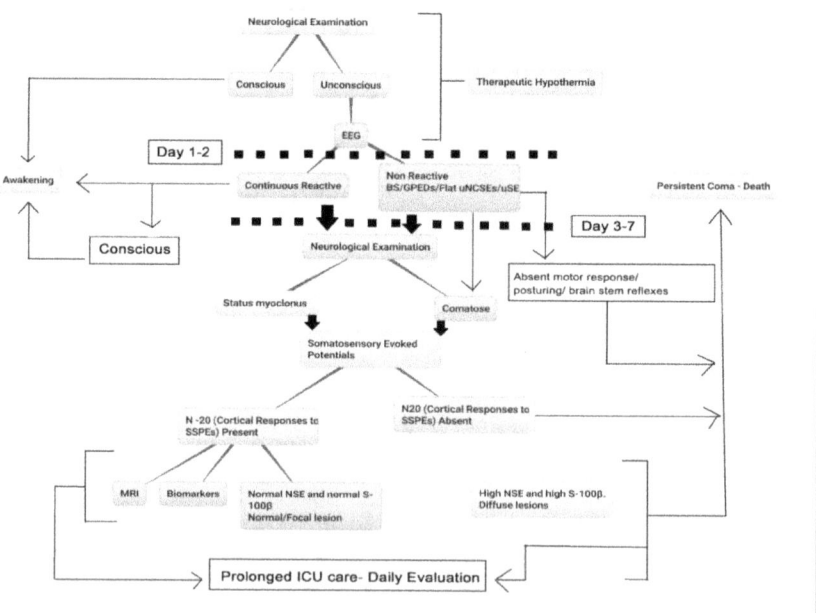

Figure: Multimodal prognostication of coma after cardiac arrest and therapeutic hypothermia. Summary of the suggested timing after cardiac arrest of all available tools that are used to predict poor outcome or neurological recovery from coma. BS, burst suppression; BSR, brainstem reflexes; EEG, electroencephalogram; GPED, generalized periodic epileptiform discharge, MRI, magnetic resonance imaging; N20, cortical responses to somatosensory evoked potentials; NSE, neuron-specific enolase; SM, status myoclonus; SSEP, somatosensory evoked potential; uNCSE, unreactive nonconvulsive seizures; uSE, unreactive status epilepticus.

7. Brain Tissue Oxygenation Monitoring

Description

Brain injury results from ischemia, tissue hypoxia, and a cascade of secondary events. Because these secondary events usually occur while patients are in the intensive care unit, the cornerstone of management includes optimization of cerebral blood flow (CBF) and oxygen and substrate delivery. New techniques for continuous monitoring of brain tissue oxygen tension (PtiO2) are now available that can provide a better understanding of complex brain physiology and help guide this management.

- The normal brain PtiO2 usually ranges from 20 to 35 mm Hg, and the ischemic threshold ranges from 10 to 19 mm Hg. Brain PtiO2 < 5 mm is associated with cell death.
- Low brain PtiO2 is associated with poor outcome and can only be detected with continuous brain PtiO2 monitoring; the sensitivity of ICP and CPP monitoring for detecting brain hypoxia is low.

Brain tissue oxygen monitoring can be done invasively or non-invasively.

Indications for Invasive Monitoring

Brain Trauma Foundation (BTF) guidelines have recommended placement of brain oxygenation monitoring when hyperventilatory strategies are employed after traumatic brain injury recommending its placement in patients at risk for ischemia.

Invasive Methods

There are two commercially available invasive probes measuring oxygen content:

- The Licox system (Integra neurosciences)
- The Neurovent-PTO system (Raumedic)

Non-invasive Methods

Near infrared spectroscopy (NIRS)

NIRS offers non-invasive online monitoring of tissue oxygenation in a wide range of clinical scenarios.

1. Cardiac Surgery:A common application is to measure cerebral oxygenation (rSO2), e.g. during cardiac surgery. Tissue hypoxia occurs frequently in the perioperative setting, particularly in cardiac surgery. Therefore, measuring and obtaining adequate tissue oxygenation may prevent (postoperative) complications and may thus be cost-effective.

2.Shock room and Trauma:NIRS may be used to detect and guide therapy in states of regional tissue hypoperfusion, even when systemic markers (e.g., blood pressure) are still within the normal range. The patho-physiological base of this discrepancy is that peripheral perfusion may be early compromised in states of hypovolemia and other forms of systemic distress, when blood volume and perfusion are redistributed towards the central compartment to protect the so-called vital organs, e.g., the heart and the brain.

Ways of Increasing Oxygen Delivery to Brain

There are following two ways of increasing oxygen delivery to brain.

1. Increasing Oxygen Delivery by Increasing CBF

Following are physiological equation of cerebral hemodynamic for management;

- Cerebral blood flow (CBF) = Cerebral Perfusion Pressure (CPP)/Cerebral Vascular Resistance (CVR)
- Cerebral Perfusion Pressure (CPP) = Mean Arterial Pressure (MAP) - Intracranial Pressure (ICP)

Summary

CBF= CPP/CVR   where as   CPP= MAP-ICP

However, acute brain injury often leads to a loss of autoregulatory mechanisms including CVR. These effects can be seen up to 2 weeks following TBI

Low brain PtiO2 can be increased by strategies aimed at increasing oxygen delivery (CBF × arterial oxygen content).The ability to improve brain PtiO2 using vasopressors to increase the MAP and CPP varies considerably depending on the status of cerebral autoregulation.Cerebral autoregulation is the ability of the brain to maintain an adequate and usually relatively constant CBF and cerebral blood volume despite changes in MAP or CPP.Increasing the CPP can improve brain PtiO2, especially when cerebral auto-regulation is impaired and the baseline PtiO2 is low.

2. Increasing Oxygen Delivery by Increasing Arterial Oxygen Content

Brain PtiO2 correlates with systemic oxygenation and PaO2.To increase PaO2 one can increase FiO2 by 100% of mechanical ventilator when baseline brain PtiO2 is high. Hyperoxia results in cerebral vasoconstriction as a way of reducing CBF in order to protect the brain from high and potentially "toxic" oxygen levels. Thus, the ability to achieve a high brain PtiO2 with 100% Fio2 may reflect impaired oxygen autoregulation. Impaired oxygen autoregulation is also associated with a poor outcome. Other way of improving PtiO2 is through increasing hemoglobin concentration. Blood transfusion may improve brain PtiO2. Whether it improves brain metabolism and leads to a cellular benefit that outweighs the risk is not clear.

The results of nonrandomized studies comparing brain PtiO2-guided therapy with ICP/CPP-guided therapy in the setting of TBI have been mixed. More studies are needed including prospective, randomized controlled trials to assess the benefit of this approach.

Note:Impaired cerebral auto-regulation and impaired oxygen autoregulation is associated with a poor outcome.

8. Imaging

Computed tomography

CT is the imaging modality of choice in the initial evaluation of patients with TBI or when acute hemorrhage is suspected. It helps to rule out surgical masses, and to identify early signs of intracranial hypertension. Since hemorrhagic lesions or edema may evolve over the first hours after injury, a CT scan must be repeated whenever there is clinical deterioration even if the initial imaging was apparently normal. New features, such as angioCT and CT perfusion, add important information to non-contrast CT and are increasingly used in stroke and SAH evaluation.

Magnetic resonance imaging

Magnetic resonance imaging (MRI) has great spatial resolution and, in patients with TBI, can identify pathologic abnormalities that are undetected or poorly characterized with CT, such as traumatic axonal injury. Acute ischemic stroke can also be detected earlier using MRI than CT. MRI is multi-parametric and can provide anatomical detail and quantitative information on brain physiology and metabolism, also allowing neuronal activation to be mapped. MRI has the advantage over other radiological methods in that it does not require ionizing radiation. Challenges for MRI scanning of critically ill patients include: monitoring and resuscitation devices that are incompatible with the magnetic field, the need for sedation or even neuromuscular blockade to prevent movement artifacts, and the risks inherent to transport outside of the ICU environment.

The Takeaway

➢ No matter how brief or extensive your neurological assessments are, comparing your findings to those of previous exams is essential.

➢ When assessing motor response, use sternal pressure judiciously. Deep sternal pressure can cause bruising.

➢ Maintain CPP in between 50-70mmHg.

➢ Normal ICP is generally in between 5–15 mmHg in healthy supine adults, 3–7 mmHg in children and 1.5–6 mmHg in infants. Maintain ICP at less than 20 to 25 mm Hg.

- The duration of monitoring is until ICP has been normal for 24 to 48 hours without ICP therapy.

- The clinical examination remains the gold standard for assessing prognosis in comatose survivors after CA; however, the use of sedatives and cooling procedures severely limit the early use of clinical findings in this setting.

- We recommend the use of a multimodal approach, including full neurological examination with at least SSEPs and EEG, to help with coma prognostication after CA and TH.

- The normal brain PtiO2 usually ranges from 20 to 35 mm Hg, and the ischemic threshold ranges from 10 to 19 mm Hg. Brain PtiO2 < 5 mm is associated with cell death.

- Low brain PtiO2 is associated with poor outcome and can only be detected with continuous brain PtiO2 monitoring; the sensitivity of ICP and CPP monitoring for detecting brain hypoxia is low.

- Maintain CPP> 60 mmHg and ICP <20 mmHg to avoid low brain PtiO2 although low brain PtiO2 can be present with CPP>60 mmHg and ICP <20mmHg.

- Impaired cerebral auto-regulation and impaired oxygen autoregulation is associated with a poor outcome.

References

1. Stocchetti N, Roux PL, Vespa P, et al. Clinical review: Neuromonitoring - an update. Critical Care. 2013;17(1):201.
2. Kellie G. Appearances observed in the dissection of two individuals; death from cold and congestion of the brain. Transactions of the Medico-Chirurgical Society of Edinburgh. 1824;1:84.
3. Magendie F. Recherches anatomique et physiologique sur le liquide céphalo-rachidien ou cérebro-spinal. Paris, France, 1842.
4. Burrows G. On Disorders of the Cerebral Circulation and on the Connection between Affections of the Brain and Diseases of the Heart. Philadelphia, Pa, USA: Lea & Blanchard; 1848.
5. Cushing H. The Third Circulation in Studies in Intracranial Physiology and Surgery. London, UK: Oxford University Press; 1926.

6. Gjerris F, Brennum J. The cerebrospinal fluid, intracranial pressure and herniation of the brain. In: Paulson OB, Gjerris F, Sørensen PS, editors. Clinical Neurology and Neurosurgery. Copenhagen, Denmark: FADL's Forlag Aktieselskab; 2004; 179–196.
7. Czosnyka M, Smielewski M, Lavinio A, Czosnyka Z, Pickard JD. A synopsis of brain pressures: which? when? Are they all useful? Neurological Research 2007;29(7):672–679.
8. Bratton SL, Chesnut RM, Ghajar J, et al. Guidelines for the management of severe traumatic brain injury. IX. Cerebral perfusion thresholds. Journal of Neurotrauma, 2007; 24: 59-64.
9. Akopian, D. J. Gaspard, and M. Alexander, "Outcomes of blunt head trauma without intracranial pressure monitoring," American Surgeon, vol. 73, no. 5, pp. 447–450, 2007.
10. Raboel PH, Bartek J, Andresen M, Bellander BM, Romner B. Intracranial Pressure Monitoring: Invasive versus Non-Invasive Methods—A Review. Critical Care Research and Practice. 2012;2012:950393.
11. Howells T, Elf K, Jones PA, Ronne-Engstrom E, Piper I, Nilsson P, et al. Pressure reactivity as a guide in the treatment of cerebral perfusion pressure in patients with brain trauma. J Neurosurg 2005;102:311-7.
12. Contant CF, Valadka AB, Gopinath SP, Hannay HJ, Robertson CS. Adult respiratory distress syndrome: A complication of induced hypertension after severe head injury. J Neurosurg 2001;95:560-8.
13. Grande PO. The Lund concept for the treatment of patients with severe traumatic brain injury. J Neurosurg Anesthesiol 2011;23:358-62.
14. Towne AR, Waterhouse EJ, Boggs JG, Garnett LK, Brown AJ, Smith Jr JR, DeLorenzo RJ. Prevalence of nonconvulsive status epilepticus in comatose patients. Neurology. 2000;54(2):340–5.
15. Rossetti AO, Reichhart MD, Schaller MD, Despland PA, Bogousslavsky J. Propofol treatment of refractory status epilepticus: a study of 31 episodes. Epilepsia. 2004;45(7):757–63.
16. Krishnamurthy KB, Drislane FW. Depth of EEG suppression and outcome in barbiturate anesthetic treatment for refractory status epilepticus. Epilepsia. 1999;40(6):759–62.
17. Claassen J, Hirsch LJ, Emerson RG, Bates JE, Thompson TB, Mayer SA. Continuous EEG monitoring and midazolam infusion for refractory nonconvulsive status epilepticus. Neurology. 2001;57(6):1036–42.
18. Arif H, Hirsch LJ. Treatment of status epilepticus. Semin Neurol. 2008;28(3):342–54.
19. Gaspard N, Hirsch LJ, LaRoche SM, Hahn CD, Westover MB, Critical Care EEGMRC. Interrater agreement for Critical Care EEG Terminology. Epilepsia. 2014;55(9):1366–73.
20. Leitinger M, Beniczky S, Rohracher A, Gardella E, Kalss G, Qerama E, Hofler J, Hess Lindberg-Larsen A, Kuchukhidze G, Dobesberger J, et al. Salzburg Consensus Criteria for Non-Convulsive Status Epilepticus—approach to clinical application. Epilepsy Behav. 2015;49:158–63.
21. Shah NA, Wusthoff CJ. How to use: amplitude-integrated EEG (aEEG). Arch Dis Child Educ Pract Ed. 2015;100(2):75–81.
22. Claassen J, Taccone FS, Horn P, Holtkamp M, Stocchetti N, Oddo M, Neurointensive Care Section of the European Society of Intensive Care M. Recommendations on the use of EEG monitoring in critically ill patients: consensus statement from the neurointensive care section of the ESICM. Intensive Care Med. 2013;39(8):1337–51.
23. Friberg H, Westhall E, Rosen I, Rundgren M, Nielsen N, Cronberg T. Clinical review: continuous and simplified electroencephalography to monitor brain recovery after cardiac arrest. Crit Care. 2013;17(4):233.
24. Karakis I, Montouris GD, Otis JA, Douglass LM, Jonas R, Velez-Ruiz N, Wilford K, Espinosa PS. A quick and reliable EEG montage for the detection of seizures in the critical care setting. J Clin Neurophysiol. 2010;27(2):100–5.
25. Young GB, Sharpe MD, Savard M, Al Thenayan E, Norton L, Davies-Schinkel C. Seizure detection with a commercially available bedside EEG monitor and the subhairline montage. Neurocrit Care. 2009;11(3):411–6.
26. Rossetti AO, Oddo M, Logroscino G, Kaplan PW. Prognostication after cardiac arrest and hypothermia: A prospective study. Ann Neurol. 2010;67:301–307.
27. Jehi LE. The Role of EEG After Cardiac Arrest and Hypothermia. Epilepsy Currents. 2013;13(4):160-161.
28. Sarkar S, Ghosh S, Ghosh SK, Collier A. Role of transcranial Doppler ultrasonography in stroke. Postgraduate Medical Journal. 2007;83(985):683-689.

29. Grundy BL. Monitoring of sensory evoked potentials during neurosurgical operations: Methods and applications. Neurosurgery 1982;11:556-75.
30. Sloan TB, Heyer EJ. Anesthesia for intraoperative neurophysiologic monitoring of the spinal cord. J Clin Neurophysiol 2002;19:430-43.
31. Cruccu G, Aminoff MJ, Curio G, Guerit JM, Kakigi R, Mauguiere F, Rossini PM, Treede RD, Garcia-Larrea L: Recommendations for the clinical use of somatosensory-evoked potentials. Clin Neurophysiol 2008, 119: 1705-1719.
32. Wijdicks EF, Hijdra A, Young GB, Bassetti CL, Wiebe S: Practice parameter: prediction of outcome in comatose survivors after cardiopulmonary resuscitation (an evidence-based review): report of the Quality Standards Subcommittee of the American Academy of Neurology. Neurology 2006, 67: 203-210.
33. Kamps MJ, Horn J, Oddo M, Fugate JE, Storm C, Cronberg T, Wijman CA, Wu O, Binnekade JM, Hoedemaekers CW: Prognostication of neurologic outcome in cardiac arrest patients after mild therapeutic hypothermia: a meta-analysis of the current literature. Intensive Care Med 2013, 39: 1671-1682.
34. Singh G. Somatosensory evoked potential monitoring. J Neuroanaesthesiol Crit Care 2016;3, Suppl S1:97-104.
35. Maas AI, Fleckenstein W, de Jong DA, van Santbrink H. Monitoring cerebral oxygenation: experimental
36. studies and preliminary clinical results of continuous monitoring of cerebrospinal fluid and brain tissue oxygen tension. Acta Neurochir Suppl (Wien). 1993;59:50–57.
37. MeixensbergerJ, Dings J, Kuhnigk H, Roosen K. Studies of tissue PO2 in normal and pathological human brain cortex. Acta Neurochir Suppl (Wien). 1993;59:58–63.
38. Zauner A, Bullock R, Di X, Young HF. Brain oxygen, CO2, pH, and temperature monitoring: evaluation in the feline brain. Neurosurgery. 1995;37(6):1168–1176; discussion 1176-1167.
39. Valadka AB, Goodman JC, Gopinath SP, Uzura M, Robertson CS. Comparison of brain tissue oxygen tension to microdialysis-based measures of cerebral ischemia in fatally head-injured humans. J Neurotrauma. 1998;15(7):509–519.
40. Meixensberger J, Kunze E, Barcsay E, Vaeth A, Roosen K. Clinical cerebral microdialysis: brain metabolism and brain tissue oxygenation after acute brain injury. Neurol Res. 2001;23(8):801–806.
41. Magnoni S, Ghisoni L, Locatelli M, . Lack of improvement in cerebral metabolism after hyperoxia in severe head injury: a microdialysis study. J Neurosurg. 2003;98(5):952–958.
42. Diringer MN, Aiyagari V, Zazulia AR, Videen TO, Powers WJ. Effect of hyperoxia on cerebral metabolic rate for oxygen measured using positron emission tomography in patients with acute severe head injury. J Neurosurg. 2007;106(4):526–529.
43. Webert KE, Blajchman MA. Transfusion-related acute lung injury. Curr Opin Hematol. 2005;12(6):480–487.
44. Marshall LF, Marshall SB, Klauber MR, van Berkum Clark M, Eisenberg HM, Jane JA, Luerssen TG, Marmarou A, Foulkes MA. A new classification of head injury based on computerized tomography. J Neurosurg. 1991;17:S14–S20.
45. Maas AI, Hukkelhoven CW, Marshall LF, Steyerberg EW. Prediction of outcome in traumatic brain injury with computed tomographic characteristics: a comparison between the computed tomographic classification and combinations of computed tomographic predictors. Neurosurgery. 2005;17:1173–1182.
46. Wijman CA, Mlynash M, Caulfield AF, Hsia AW, Eyngorn I, Bammer R, Fischbein N, Albers GW, Moseley M. Prognostic value of brain diffusion-weighted imaging after cardiac arrest. Ann Neurol. 2009;17:394–402.
47. Sidaros A, Engberg AW, Sidaros K, Liptrot MG, Herning M, Petersen P, Paulson OB, Jernigan TL, Rostrup E. Diffusion tensor imaging during recovery from severe traumatic brain injury and relation to clinical outcome: a longitudinal study. Brain. 2008;17:559–572.
48. Tollard E, Galanaud D, Perlbarg V, Sanchez-Pena P, Le Fur Y, Abdennour L, Cozzone P, Lehericy S, Chiras J, Puybasset L. Experience of diffusion tensor imaging and 1H spectroscopy for outcome prediction in severe traumatic brain injury: Preliminary results. Crit Care Med. 2009;17:1448–1455.
49. Tong KA, Ashwal S, Holshouser BA, Nickerson JP, Wall CJ, Shutter LA, Osterdock RJ, Haacke EM, Kido D. Diffuse axonal injury in children: clinical correlation with hemorrhagic lesions. Ann Neurol. 2004;17:36–50.

50. Kasahara M, Menon DK, Salmond CH, Outtrim JG, Taylor Tavares JV, Carpenter TA, Pickard JD, Sahakian BJ, Stamatakis EA. Altered functional connectivity in the motor network after traumatic brain injury. Neurology. 2010;17:168–176.

Hemodynamic monitoring in the critically ill: an overview of current cardiac output monitoring methods

Salman Assad, MD

Department of Neurology and Neurosurgery

Shifa Tameer-e-Millat University, Islamabad, Pakistan

Introduction

Patients admitted to the intensive care unit (ICU) in general suffer from organ failure (single or multiple) or are at risk of such organ failure, which includes patients after major surgery and/or trauma. Hemodynamic instability, causing a mismatch between oxygen delivery and demand, is a major contributive factor for organ failure. Alterations in effective circulating volume (e.g. hypovolemia), cardiac function, and/or vascular tone (e.g. vasoplegic shock in sepsis) underlie hemodynamic instability [1]. We can often manage it with regular clinical examination and monitoring of certain basic vital parameters (heart rate, blood pressure, central venous pressure [CVP], peripheral and central venous oxygen saturation, and respiratory variables) and urine output, but when these fail there is an increased need for hemodynamic monitoring (cardiac output [CO], pulmonary arterial occlusion pressure [PAOP or wedge pressure], pulmonary arterial pressure [PAP], mixed venous oxygen saturation [SvO_2], stroke volume variation [SVV], extravascular water, etc.) to guide fluid management and vasopressor/inotropic support. Over the last few decades, hemodynamic monitoring has evolved from basic monitoring of CO to sophisticated devices providing a plethora of variables. These techniques and devices can be classified in either of two ways: 1) calibrated versus non-calibrated techniques and 2) by their degree of invasiveness (invasive, less invasive, or non-invasive). In this article, we will provide an overview of the indications and limitations for hemodynamic monitoring and the available methods of doing so.

Indications for hemodynamic monitoring

All patients admitted to the ICU should be monitored, but the degree of monitoring can vary. Hemodynamically stable patients require maybe nothing more than continuous electrocardiographic (ECG) monitoring, regular non-invasive blood pressure measurement, and peripheral pulse oximetry (peripheral oxygen saturation or SpO_2). Those who are unstable, or at risk of instability, should receive an arterial line for continuous invasive blood pressure measurement and regular analysis of arterial blood gasses. Any patient receiving vasopressors or inotropic agents requires a central venous line for drug administration and, when indicated, measurement of CVP and central venous oxygen saturation ($ScvO_2$). When initial resuscitation fails to improve the hemodynamic and/or respiratory status of the patient, advanced hemodynamic monitoring will be required to guide medical management. Measuring CO and its components (preload, afterload, and contractility) will tell us if there is ongoing need for fluid resuscitation, vasopressors, or inotropic agents. It can be used as a diagnostic tool to determine the type of shock (hypovolemic, cardiogenic, obstructive, or distributive) according to the hemodynamic profile. Furthermore, it can be used to guide de-resuscitation, the phase after re-convalescence during which we are often confronted with fluid overload (in itself an important negative prognostic predictor) [2,3]. The clinical context (emergency room, operating room, or ICU) and the different possible variables provided by the monitoring method will determine which method we will use. There is, however, an important remark to be added when discussing indications for monitoring. Trials have as of yet not been able to show a significant reduction in mortality when comparing monitoring to standard of care, although there are possible benefits concerning complications [4-7]

Basics of hemodynamic monitoring

Measuring the CO starts with understanding the Fick principle, described by Adolf Fick in 1870 [8]. In essence, this states that the blood flow to an organ can be calculated by using an indicator and measuring the amount of indicator that is taken up by the organ and its respective concentrations in arterial and venous blood. When we think of the entire human body as the organ described and use oxygen as the indicator, we can measure CO using this formula:

$$CO = VO_2 CaO_2 - CvO_2$$

In this formula, VO_2 is the consumption of oxygen and CaO_2 and CvO_2 are the arterial and mixed venous oxygen contents, respectively. The VO_2 can be measured using a spirometer within a closed rebreathing circuit. Arterial and mixed venous oxygen are measured using blood samples from a peripheral arterial line (oxygenated blood) and a pulmonary artery catheter (PAC) (deoxygenated blood), respectively. This method is therefore invasive and time consuming, and although considered the gold standard it is rarely performed.

Methods of hemodynamic monitoring

Several invasive and less-invasive methods have been developed during the last few decades to measure CO. The first to be used was the PAC, introduced in the 1970's by Swan, Ganz, and Forrester [2]. It is still the gold standard in the clinical setting to which we refer when comparing different methods of hemodynamic monitoring. These can be classified as calibrated or non-calibrated techniques or according to their level of invasiveness (invasive, less invasive, or non-invasive). There is a trend to use more less-invasive and non-invasive techniques to reduce the risks that accompany (less) invasive techniques.

Repeated calibration is performed in order to eliminate or reduce bias in continuous measurements. It refers to the act of evaluating and adjusting the precision and accuracy of the equipment. The precision of a technique is the degree to which repeated measurements (at the same time) show the same results, and the accuracy is the degree of closeness of the results to the actual true value (obtained by the gold standard method). Non-calibrated techniques try to reduce bias by implementing correction factors based on patient demographics (age, weight, gender, etc.) or calculations. However, in situations where preload, afterload, contractility, and aortic compliance can vary widely (as in critical illness), calibration will often prove necessary.

Invasive techniques

Pulmonary artery catheter (calibrated). The gold standard, the PAC, is a flow-directed catheter that is placed through an introducer in the jugular, subclavian, or, more seldom, the femoral vein and that travels from the right atrium through the right ventricle just until the pulmonary artery. It allows direct simultaneous measurement of pressures in the right atrium (CVP), PAP, and PAOP or wedge pressure, which in turn is indicative of the filling pressures in the left atrium. Blood sampling from the distal port (pulmonary artery) allows measurement of SvO_2, and using fiber optic reflectometry allows for continuous monitoring of the SvO_2. CO is measured with thermodilution. Initially, a cold saline bolus has to be delivered through the opening in the right atrium, with a thermistor detecting the drop in temperature a few centimeters from the tip of the catheter. Later, a heating coil is incorporated in the design, negating the need for cold fluid boluses (and thus avoiding bias because of different operators). This CO measurement, however, is not a true continuous monitoring seeing as it represents the average value of the last 5 minutes, and changes in CO during alterations in preload or afterload (e.g. fluid challenge) cannot be appreciated instantaneously. It also provides several calculated variables such as systemic and pulmonary vascular resistance, left and right ventricular stroke work, and the oxygen extraction ratio. Intracardiac electrodes allow the monitoring of electric activity, from which volumetric variables such as right ventricular ejection fraction (RVEF) and continuous assessment of right ventricular end diastolic volume (CEDV) can be gauged, providing information concerning right ventricular contractility and preload, respectively.

Although PAC was the most widely used technique in the past, a clear survival benefit has not been proven[10]. The complexity of possible variations in obtained pressure tracings has led to large inter-observer variability, together with reports of very common misinterpretation of tracings [11].

The best indication for the PAC remains when there is right ventricular heart failure or pulmonary hypertension, seeing as no other monitoring device is capable of providing direct measurement of the pressures in the right heart and pulmonary circulation.

Less-invasive techniques

1. Transpulmonary thermodilution: the PiCCO ® system (calibrated/surrogate gold standard).

Using a central venous catheter and arterial line with thermistor, the PiCCO ® system provides both intermittent (for calibration) and continuous CO measurement. The intermittent CO is measured using a transpulmonary thermodilution technique, where a cold fluid bolus is injected through the central line. Using the Stewart Hamilton equation, the area under the thermodilution curve is then used to calculate the CO. By using an algorithm based on the analysis of the arterial pulse contour, it is possible to continuously monitor CO and stroke volume, allowing assessment of beat-to-beat variations of stroke volume and CO in changing preload conditions. SVV and pulse pressure variation (PPV) have been proposed as variables to guide fluid loading in critical care settings [12, 13], although limited to completely sedated patients under controlled mechanical ventilation and in the absence of cardiac arrhythmias (LIMITS: low heart rate/respiratory rate ratio, irregular heart beats, mechanical ventilation with low tidal volume, increased abdominal pressure, thorax open, spontaneous breathing) [14].

Furthermore, the PiCCO ® system allows the measurement of global end diastolic volume (GEDV), intrathoracic blood volume (ITBV), and extravascular lung water (EVLW). Pulmonary blood volume (PBV), pulmonary vascular permeability index (PVPI), global ejection fraction (GEF), contractility, and systemic vascular resistance (SVR) are derived from these values. These values can be indexed to body surface area and predicted body weight.

This system has several advantages over PAC: it is less invasive, it provides a true continuous CO and rapidly available measurements allowing the assessment of fluid responsiveness, and it is supported by literature data in humans that show good correlation between intermittent and continuous transpulmonary thermodilution CO with the PAC as gold standard.

Its drawbacks are the need for a specialized arterial line (typically placed in the femoral artery), a central venous line (jugular or subclavian vein), and regular calibration (three to four times a day) with cold fluid boluses (extra fluid load). The volume measurement is not automated and not continuous. It is less useful in valvulopathies, abdominal aortic aneurysm, or enlarged atria, and it is not applicable in arrhythmias or during intra-aortic balloon counterpulsation.

2. Transpulmonary thermodilution: the VolumeView ®/EV1000 ® system (calibrated). The VolumeView®/EV1000 ® system is similar to the PiCCO ® system but differs in the measurement of the GEDV, where it uses a formula implementing the maximum upslope and downslope time of the thermodilution curve, whereas the PiCCO ® system employs time constants derived from the mean appearance, mean transit, and downslope of the thermodilution curve [15].

3. Transpulmonary dye dilution: the LiDCO ® system (calibrated). Instead of thermal dilution, the LiDCO ® system uses lithium as an intravascular indicator injected through a central or peripheral vein which is then measured in a peripheral artery using a specialized sensor probe attached to the pressure line[16]. It is coupled to a pulse contour analysis system (LiDCOrapid ®/PulseCO ®). The only additional measured variables compared to PAC monitoring are the PPV and SVV. The data are rapidly available and provide real-time beat-to-beat variations in CO. Volume quantification, however, is not available, and the technique cannot be used in children/patients with a weight below 40 kg or patients under the influence of muscle relaxants (the positively charged quaternary ammonia ion is detected by the lithium sensor, affecting its measurements). Little is known about possible toxic effects or accumulation with long-term use of lithium. Furthermore, the ion-selective electrode is delicate and expensive and needs to be replaced every three days.

4. Ultrasound flow dilution: the COstatus ® system (calibrated). The COstatus ® system calculates CO by using transpulmonary ultrasound dilution technology to measure changes in blood ultrasound velocity and blood flow following an injection of saline [17]. It requires a primed extracorporeal arteriovenous tube set (AV loop) connected between the *in situ* standard arterial catheter and central venous catheter where two ultrasound flow-dilution sensors are placed on the arterial and venous ends. During calibration, a small roller pump is used to circulate blood through the AV loop from the artery to the vein. The ultrasound sensors provide an ultrasound dilution curve through which CO can be calculated following the Stewart Hamilton principle. After calibration, a continuous CO can be calculated through the arterial waveform. It calculates certain volumetric indices such as total end diastolic volume (TEDV), central blood volume (CBV), and active circulation volume (ACV), and it can detect intracardiac shunts. It is validated in both adult and pediatric patients. Recalibration is necessary in unstable conditions.

5. Pulse contour and pulse pressure analysis (non-calibrated). Several devices use the technique of pulse pressure analysis to estimate CO. The difficulty is that, to estimate CO from pulse pressure analysis, one would not only need information about the heart rate and blood pressure but also have to make an estimate about the pressure-volume relationship of the aorta. Most of the techniques being used today are based on a three-element model integrating aortic characteristic impedance, arterial compliance, and systemic vascular resistance. These models work relatively well in stable patients but lack accuracy in unstable patients or when vasoactive drugs are employed [18]. There are several devices using pulse pressure analysis available:

- FloTrac ®/Vigileo ®: a widely used method that uses PPV and vascular tone to calculate stroke volume and CO, although it is less useful in situations with low vascular tone (e.g. septic shock) [19].

- ProAQT ®/Pulsioflex ®: continuously measures CO by analyzing the systolic portion of the pressure wave after an initial autocalibration (depending on patient characteristics) or manually entering a starting cardiac index; there is, however, too large a percentage error [20].

- LiDCOrapid ®/pulseCO ®: uses the same algorithm as in LiDCOplus with calculating a nominal stroke volume from the entire pressure waveform. It can be calibrated using other techniques. There is, however, insufficient accuracy compared with thermodilution methods 21. Calibration improved this accuracy (even in critically ill patients) but only for the first four hours 22.

- Most Care ®/pressure recording analytical method (PRAM): uses an algorithm called "pressure recording analytical method", which is a theoretical method developed by analyzing both pulsatile and continuous flow 23; only an invasive arterial catheter is needed, the implementation is easy, and it shows good accuracy in a wide range of settings.

6. Respiratory derived cardiac output monitoring system: partial CO_2-rebreathing (NiCO ®) *(non-calibrated).* Using CO_2 instead of O_2 as an indicator in the Fick principle (see above), the NiCO ® uses a partial rebreathing method to measure the CO. The system consists of a CO_2 and airflow sensor combined with a pulse oximeter. We can measure the CO_2 production by multiplying the exhaled CO_2 content by the respiratory minute volume. The arterial CO_2 is derived from the end tidal CO_2. Every three minutes, a partial rebreathing cycle should be started using a rebreathing loop, resulting in reduced CO_2 elimination. By assuming CO is stable in both normal and rebreathing conditions, the difference between normal and rebreathing ratios are used to calculate CO. However, as it is dependent on stable ventilation, this can be used only in fully sedated patients with volume-controlled ventilation. Significant pulmonary disease (as in ICU patients with acute respiratory distress syndrome, pneumonia, atelectasis, shunting, etc.) can interfere with the measurements. To date, insufficient data exist to support its accuracy, specifically in critically ill patients.

7. Transesophageal echocardiography (operator dependent). Transesophageal echocardiography (TEE) is an important cardiovascular diagnostic tool in perioperative and critical care medicine. It uses ultrasound to provide real-time images of the cardiac structures and blood flow. The transducer is placed in the esophagus next to the heart to produce these images. It may help define pathophysiological abnormalities in patients like wall motion abnormalities, pericardial effusions, pulmonary hypertension, and valvulopathy, in conjunction with other invasive or less-invasive monitoring. Guidelines published by the American Society of Anesthesiologists and the Society of Cardiovascular Anesthesiologists state that TEE should be used in critical care patients with persistent hypotension or hypoxia when diagnostic information expected to alter management cannot be obtained by transthoracic echocardiography (TTE) or other modalities in a timely manner [24]. There is, however, a significant learning curve, TEE is expensive, and continuous monitoring is not an option. There is a (low) risk of oropharyngeal bleeding and dislocation of the endotracheal tube, and its use is relatively contraindicated in esophageal pathologies and severe coagulation abnormalities.

8. Esophageal Doppler (operator dependent). Using a flexible ultrasound probe, the blood flow in the descending aorta is measured to determine stroke volume and CO. This probe can be left in place for prolonged periods of time (barring dislocation) and can provide real-time CO as well as afterload data interpretation. It provides many additional measurements as well as an estimate for preload via the corrected flow time. It is a promising, easy-to-learn technique associated with reduced hospital stay and better perioperative volume optimization [25].

Non-invasive techniques

1. Transthoracic echocardiography (operator dependent). CO can be measured with TTE using pulsed wave Doppler velocity in the left ventricular outflow tract (LVOT). It can also be measured at the mitral valve annulus, ascending aorta, right ventricular outflow tract (RVOT), and pulmonary artery, but these have been less validated. Seeing as there is less influence of systemic vascular resistance (SVR), measurements over the RVOT can provide an accurate CO, but only if there is no interference due to pulmonary arterial hypertension.

2. Non-invasive pulse contour systems (non-calibrated). These systems strive to determine CO based on an arterial pulse pressure curve, which is estimated by a completely non-invasive technique.

3. Bioimpedance (non-calibrated). Using skin electrodes, a small electrical current is applied. Changes in voltage over the circuit are then caused by changes in impedance and/or volume of the conducting tissues. Blood has a relatively low resistivity, and changes in intrathoracic blood volume have a high impact on impedance accordingly. With this assumption, we can postulate that changes in thoracic impedance are largely dependent on three components: a baseline impedance indirectly proportional to the thoracic fluid content, tidal changes in intrathoracic blood volume caused by respiration, and small changes caused by the cardiac cycle. The latter are primarily due to changes in aortic volume, which can be used to estimate stroke volume and CO [29, 30]. However, it does have important limitations. The impedance is influenced by all changes in thoracic fluid composition such as lung edema and pleural effusions. Changes in systemic vascular resistance will influence the volume changes in the aorta and will therefore interfere with CO measurements.

4. Estimated continuous cardiac output (esCCO ®) (non-calibrated). This is a non-invasive device estimating the CO with an algorithm based on patient characteristics and measurement of heart rate, peripheral oxygen saturation, and non-invasive blood pressure. With these measurements, a pulse wave transit time is determined and combined with the heart rate to estimate the CO. Although it has the advantage of being non-invasive, it remains a mere estimation of the CO. Studies suggest an unacceptable high deviation compared to validated methods [31, 32].

5. Ultrasonic cardiac output monitoring (USCOM ®) (non-calibrated). Measuring the flow velocity in the aortic and pulmonary outflow tracts, USCOM ® combines this with pre-calculated valve areas to estimate a CO. It has a short learning curve and has few procedural risks. There is, however, quite a proportion of unobtainable imaging, the proposed valve areas can differ significantly from the truth (specifically in elderly patients, patients who are critically ill, and patients with structural heart disease), and there can be a big difference between the estimated output and the calibrated reference value [33–36].

Conclusion

Critically ill patients are often hemodynamically unstable (or at risk of becoming unstable), and advanced hemodynamic monitoring is recommended in complex situations or in patients with shock who do not respond to initial fluid resuscitation. We are offered a wide variety of techniques that range from invasive to less invasive and even non-invasive. These techniques can be calibrated or non-calibrated. In Table 1, a schematic overview is given of the discussed techniques with their respective advantages and disadvantages. Calibrated techniques offer the best precision and accuracy, and the obtained values concerning CO, preload, afterload, and different other derived values are of significant value in the hemodynamic stabilization of critically ill patients. Relying on non-calibrated techniques can prove difficult in critically ill patients, where rapidly changing conditions in preload, vasomotor tone, and cardiac function can often lead to misleading results, with a risk of inappropriate medical management, under- or over-resuscitation, and subsequent organ dysfunction. They can be of value, however, in stable conditions, with less- or non-invasive techniques negating the possibility of complications due to more invasive techniques. Pulse contour analysis, in particular, with the added functional variables SVV and PPV, can be of significant value in the assumption that the patient is in regular sinus rhythm and fully sedated under controlled mechanical ventilation. As is so often required in the medical management of critically ill patients, we will have to balance the benefits and risks of the different techniques in the hope of achieving the best possible outcome for our patient. We recommend using calibrated techniques in the critically ill and unstable patients, preferring less-invasive techniques to more-invasive ones. A PAC, however, can be particularly useful in patients with significant cardiac dysfunction, specifically when concerning right ventricular dysfunction or pulmonary arterial hypertension. During de-resuscitation, the monitoring technique should be re-evaluated (and likewise when the patient deteriorates again), and non-invasive techniques should be used whenever possible instead of (less) invasive techniques. Non-invasive techniques can be combined with transthoracic/transesophageal echocardiography to provide valuable additional information.

Table 1: Overview of monitoring methods.

Method	Examples of commercial name	Calibrated or not	Major advantages	Major disadvantages
Invasive methods				
Pulmonary artery catheter		Calibrated	Direct measurements in right atrium and pulmonary circulation	Delay in determining CO, most invasive, and risks involved
Less-invasive methods				
Transpulmonary thermodilution	PiCCO ® VolumeView ®/EV1000® LiDCO ®	Calibrated	Intermittent and continuous CO, added variables	Need for specialized arterial and central venous line, LIMITS (PiCCO ® system)
Ultrasound flow dilution	COstatus ®	Calibrated	Continuous CO, added variables, can	Requires AV loop

			detect intracardiac shunts	
Pulse contour and pulse pressure variation	FloTrac ®/Vigileo ® ProAQT ®/Pulsioflex ® LiDCOrapid ®/pulseCO® Most Care ®/PRAM	Non-calibrated	Continuous CO	Lack accuracy in unstable patients or during use of vasoactive drugs
Partial CO_2 -rebreathing	NiCO ®	Non-calibrated	No need for intravascular devices	Only in sedated patients under volume control ventilation, interference from pulmonary disease
Transesophageal echocardiography		Operator dependent	Real-time images of the cardiac structures and blood flow	Learning curve, (low) risk of complications
Esophageal Doppler		Operator dependent	Real-time CO and afterload data, added variables	Risk of dislocation

Non-invasive methods

Transthoracic echocardiography		Operator dependent	Direct measurement of CO and visualization of cardiac structures	Ultrasound characteristics often suboptimal in ICU patients
Non-invasive pulse contour systems	T-line ® ClearSight ®/Nexfin ®/ Physiocal ® CNAP ®/VERIFY ®	Non-calibrated	Non-invasive, simple tool	Less accurate, needs more validation
Bioimpedance		Non-calibrated	Simple tool, providing data concerning CO and fluid overload	Changes intrathoracic fluid content and SVR influence measurements
Estimated continuous cardiac output ®	esCCO ®	Non-calibrated	Uses widely available variable to estimate CO	Is only estimate, inadequate accuracy

| Ultrasonic cardiac output monitoring ® | USCOM ® | Non-calibrated | Short learning curve and only few risks | Only estimate, uses standard valve areas which can differ in patients |

AV loop, arteriovenous fistula; CO, cardiac output; ICU, intensive care unit; SVR, systemic vascular resistance.

References

1. Teboul JL, Saugel B, Cecconi M, et al. : Less invasive hemodynamic monitoring in critically ill patients. Intensive Care Med. 2016;42(9):1350–9. 10.1007/s00134-016-4375-7.

2. Malbrain ML, Marik PE, Witters I, et al. : Fluid overload, de-resuscitation, and outcomes in critically ill or injured patients: a systematic review with suggestions for clinical practice. Anaesthesiol Intensive Ther. 2014;46(5):361–80. 10.5603/AIT.2014.0060 [PubMed] [Cross Ref]

3. Vincent JL, Sakr Y, Sprung CL, et al. : Sepsis in European intensive care units: results of the SOAP study. Crit Care Med. 2006;34(2):344–53. 10.1097/01.CCM.0000194725.48928.3A.

4. ProCESS Investigators, . Yealy DM, Kellum JA, et al. : A randomized trial of protocol-based care for early septic shock. N Engl J Med. 2014;370(18):1683–93. 10.1056/NEJMoa1401602.

5. ARISE Investigators, . ANZICS Clinical Trials Group, . Peake SL, et al. : Goal-directed resuscitation for patients with early septic shock. N Engl J Med. 2014;371(16):1496–506. 10.1056/NEJMoa1404380 [PubMed] [Cross Ref]

6. Mouncey PR, Osborn TM, Power GS, et al. : Trial of early, goal-directed resuscitation for septic shock. N Engl J Med. 2015;372(14):1301–11. 10.1056/NEJMoa1500896.

7. Pearse RM, Harrison DA, MacDonald N, et al. : Effect of a perioperative, cardiac output-guided hemodynamic therapy algorithm on outcomes following major gastrointestinal surgery: a randomized clinical trial and systematic review. JAMA. 2014;311(21):2181–90. 10.1001/jama.2014.5305.

8. Fick A: Uber die messing des Blutquantums in den Hertzvent rikeln.Sitzber Physik Med Ges Wurzburg.1870;36.

9. Swan HJ, Ganz W, Forrester J, et al. : Catheterization of the heart in man with use of a flow-directed balloon-tipped catheter. N Engl J Med. 1970;283(9):447–51. 10.1056/NEJM197008272830902.

10. Sandham JD, Hull RD, Brant RF, et al. : A randomized, controlled trial of the use of pulmonary-artery catheters in high-risk surgical patients. N Engl J Med. 2003;348(1):5–14. 10.1056/NEJMoa021108.

11. Squara P, Bennett D, Perret C: Pulmonary artery catheter: does the problem lie in the users? Chest. 2002;121(6):2009–15. 10.1378/chest.121.6.2009.

12. Goedje O, Hoeke K, Lichtwarck-Aschoff M, et al. : Continuous cardiac output by femoral arterial thermodilution calibrated pulse contour analysis: comparison with pulmonary arterial thermodilution. Crit Care Med. 1999;27(11):2407–12. [PubMed]

13. Michard F, Teboul JL: Predicting fluid responsiveness in ICU patients: a critical analysis of the evidence. Chest. 2002;121(6):2000–8. 10.1378/chest.121.6.2000.

14. Michard F, Chemla D, Teboul JL: Applicability of pulse pressure variation: how many shades of grey? Crit Care. 2015;19:144. 10.1186/s13054-015-0869-x.

15. Kiefer N, Hofer CK, Marx G, et al. : Clinical validation of a new thermodilution system for the assessment of cardiac output and volumetric parameters. Crit Care. 2012;16(3):R98. 10.1186/cc11366.

16. Jonas MM, Tanser SJ: Lithium dilution measurement of cardiac output and arterial pulse waveform analysis: an indicator dilution calibrated beat-by-beat system for continuous estimation of cardiac output. Curr Opin Crit Care. 2002;8(3):257–61. 10.1097/00075198-200206000-00010.

17. Galstyan G, Bychinin M, Alexanyan M, et al. : Comparison of cardiac output and blood volumes in intrathoracic compartments measured by ultrasound dilution and transpulmonary thermodilution methods. Intensive Care Med. 2010;36(12):2140–4. 10.1007/s00134-010-2003-5.

18. Cecconi M, Malbrain ML: Cardiac output obtained by pulse pressure analysis: to calibrate or not to calibrate may not be the only question when used properly. Intensive Care Med. 2013;39(4):787–9. 10.1007/s00134-012-2802-y [PubMed] [Cross Ref]

19. Marqué S, Gros A, Chimot L, et al. : Cardiac output monitoring in septic shock: evaluation of the third-generation Flotrac-Vigileo. J Clin Monit Comput. 2013;27(3):273–9. 10.1007/s10877-013-9431-z.

20. Monnet X, Vaquer S, Anguel N, et al. : Comparison of pulse contour analysis by Pulsioflex and Vigileo to measure and track changes of cardiac output in critically ill patients. Br J Anaesth. 2015;114(2):235–43. 10.1093/bja/aeu375.

21. Phan TD, Kluger R, Wan C, et al. : A comparison of three minimally invasive cardiac output devices with thermodilution in elective cardiac surgery. Anaesth Intensive Care. 2011;39(6):1014–21.

22. Cecconi M, Fawcett J, Grounds RM, et al. : A Prospective Study to Evaluate the Accuracy of Pulse Power Analysis to Monitor Cardiac Output in Critically Ill Patients. BMC Anesthesiol. 2008;8:3. 10.1186/1471-2253-8-3.

23. Scolletta S, Romano SM, Biagioli B, et al. : Pressure recording analytical method (PRAM) for measurement of cardiac output during various haemodynamic states. Br J Anaesth. 2005;95(2):159–65. 10.1093/bja/aei154.

24. American Society of Anesthesiologists and Society of Cardiovascular Anesthesiologists Task Force on Transesophageal Echocardiography: Practice guidelines for perioperative transesophageal echocardiography. An updated report by the American Society of Anesthesiologists and the Society of Cardiovascular Anesthesiologists Task Force on Transesophageal Echocardiography. Anesthesiology. 2010;112(5):1084–96. 10.1097/ALN.0b013e3181c51e90.

25. McKendry M, McGloin H, Saberi D, et al. : Randomised controlled trial assessing the impact of a nurse delivered, flow monitored protocol for optimisation of circulatory status after cardiac surgery. BMJ. 2004;329(7460):258. 10.1136/bmj.38156.767118.7C.

26. Saugel B, Meidert AS, Langwieser N, et al. : An autocalibrating algorithm for non-invasive cardiac output determination based on the analysis of an arterial pressure waveform recorded with radial artery applanation tonometry: a proof of concept pilot analysis. J Clin Monit Comput. 2014;28(4):357–62. 10.1007/s10877-013-9540-8.

27. Ameloot K, Palmers PJ, Malbrain ML, et al. : The accuracy of noninvasive cardiac output and pressure measurements with finger cuff: a concise review. Curr Opin Crit Care. 2015;21(3):232–9. 10.1097/MCC.0000000000000198.

28. Wagner JY, Grond J, Fortin J, et al.: Continuous noninvasive cardiac output determination using the CNAP system: evaluation of a cardiac output algorithm for the analysis of volume clamp method-derived pulse contour. J Clin Monit Comput. 2016;30(4):487–93. 10.1007/s10877-015-9744-1.

29. Summers RL, Shoemaker WC, Peacock WF, et al. : Bench to bedside: electrophysiologic and clinical principles of noninvasive hemodynamic monitoring using impedance cardiography. Acad Emerg Med. 2003;10(6):669–80. 10.1111/j.1553-2712.2003.tb00054.x.

30. Moshkovitz Y, Kaluski E, Milo O, et al. : Recent developments in cardiac output determination by bioimpedance: comparison with invasive cardiac output and potential cardiovascular applications. Curr Opin Cardiol. 2004;19(3):229–37. 10.1097/00001573-200405000-00008.

31. Ball TR, Tricinella AP, Kimbrough BA, et al. : Accuracy of noninvasive estimated continuous cardiac output (esCCO) compared to thermodilution cardiac output: a pilot study in cardiac patients. J Cardiothorac Vasc Anesth. 2013;27(6):1128–32. 10.1053/j.jvca.2013.02.019.

32. Sinha AC, Singh PM, Grewal N, et al. : Comparison between continuous non-invasive estimated cardiac output by pulse wave transit time and thermodilution method. Ann Card Anaesth. 2014;17(4):273–7. 10.4103/0971-9784.142059.

33. Van den Oever HL, Murphy EJ, Christie-Taylor GA: USCOM (Ultrasonic Cardiac Output Monitors) lacks agreement with thermodilution cardiac output and transoesophageal echocardiography valve measurements. Anaesth Intensive Care. 2007;35(6):903–10. [PubMed]

34. Thom O, Taylor DM, Wolfe RE, et al. : Comparison of a supra-sternal cardiac output monitor (USCOM) with the pulmonary artery catheter. Br J Anaesth. 2009;103(6):800–4. 10.1093/bja/aep296.

35. Boyle M, Steel L, Flynn GM, et al. : Assessment of the clinical utility of an ultrasonic monitor of cardiac output (the USCOM) and agreement with thermodilution measurement. Crit Care Resusc. 2009;11(3):198–203.

36. Nguyen HB, Banta DP, Stewart G, et al. : Cardiac index measurements by transcutaneous Doppler ultrasound and transthoracic echocardiography in adult and pediatric emergency patients. J Clin Monit Comput. 2010;24(3):237–47. 10.1007/s10877-010-9240-6.

www.ingramcontent.com/pod-product-compliance
Lightning Source LLC
Chambersburg PA
CBHW070948200526
45161CB00001BA/26